SPACE GUIDES

EXPLORING THE MOON

PETER GREGO

QED Publishing

Copyright © QED Publishing 2007

First published in the UK in 2007 by
QED Publishing
A Quarto Group company
226 City Road
London EC1V 2TT
www.qed-publishing.co.uk

A catalogue record for this book is available from the British Library.

ISBN 978-1-84538-685-6

Written by Peter Grego
Produced by Calcium
Editor Sarah Eason
Illustrations by Geoff Ward
Picture Researcher Maria Joannou

Publisher Steve Evans
Creative Director Zeta Davies
Senior Editor Hannah Ray

Printed and bound in China

Picture credits

Key: T = top, B = bottom, C = centre, L = left, R = right, FC = front cover, BC = back cover

Corbis/Rob Matheson 6T, /Roger Ressmeyer 5B, /Ute & Juergen Schimmelpfennig/Zefa 9B, /Doug Wilson 18–19; **Getty Images**/Photodisc 1, 4, 10–11, 12, /Stone 6B; **istockphoto. com**/Nico Smit 19; **Jamie Cooper** 17T; **NASA** FCT, FCB, 11B, 14, 15B, 15T, 16B, 17B, 20, 21T, 22, 23B, 23T, 24, 25T, 27T, 28, 29B, 29T, BC, /Neil A. Armstrong 25B, / Charles M. Duke Jr 27B, /Peter Grego 7T, /JPL 8, /JPL-Caltech 9T, /Kennedy Space Center 21B, /Harrison H. Schmitt FCC, 26; **NSSDC** 3; **Rex Features**/Everett Collection 18; **Science Photo Library**/Russell Croman 13, 16T, /John Sanford 11T, /Kaj R Svensonn 7B

Words in **bold** can be found in the Glossary on pages 30–31.

Contents

The lonely Moon

The **Moon** is the Earth's only natural **satellite**. It is a solid ball of rock, about as wide as the United States. Although the Moon is our nearest neighbour in **space**, it would take you about six months to reach it from Earth on a non-stop, high-speed train. Astronauts have reached the Moon in just three days – but they were hurtling through space at thousands of kilometres per hour.

Amazing

Bigger than Pluto

Our Moon is bigger than the dwarf **planet** Pluto! It is the fourth-biggest moon in the Solar System.

The Moon is 384 400km away from Earth – the same distance you would cover if you travelled around the planet 10 times. ⇧

4

The lunar month

The Moon takes about a month to make one **orbit** around the Earth. It keeps the same side turned towards us, so there is a far side of the Moon that we never see.

The Moon does not shine – it has no light of its own. Its surface is illuminated by light from the **Sun**. In the course of a month, the Moon appears to change shape as it orbits the Earth, in a series of what are called **phases**. At the start of the **lunar** month, the Moon looks like a **crescent** in the sky. A week later the half-Moon appears, and after another week it has become a complete circle. It then narrows to the half-Moon again, before becoming a thinning crescent. However, the Moon is not actually changing shape. As it orbits the Earth, we are just seeing different areas that have come into the sunlight and are being lit up.

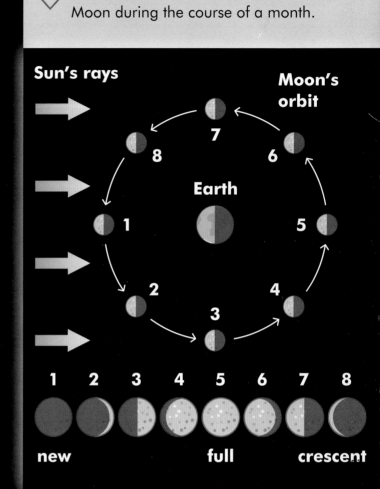

This diagram shows the phases of the Moon during the course of a month.

Earthshine

When the Moon is a crescent, we may see the rest of it lit with a faint, bluish glow. This is called **earthshine**, and is caused by light being reflected on to the Moon from the Earth.

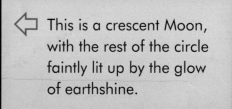

This is a crescent Moon, with the rest of the circle faintly lit up by the glow of earthshine.

The Moon's myths and legends

Throughout history, people around the world have enjoyed including the Moon in their myths and legends.

The crescent Moon was once thought to be the horns of a magical bull in the sky.

A mighty bull

People used to imagine that the Moon had magical powers, because it appears to change shape. One Ancient Egyptian legend says that the crescent Moon is the horns of a once-mighty bull that was placed in the sky when it died on Earth. The bull comes back to life in the heavens when it appears to grow wider, becoming its most powerful at **full Moon**.

Creatures in the Moon

At full Moon you can see a patchwork of dark and light areas on the Moon's surface. With a bit of imagination, it is possible to make out strange faces and creatures in these dark and light areas.

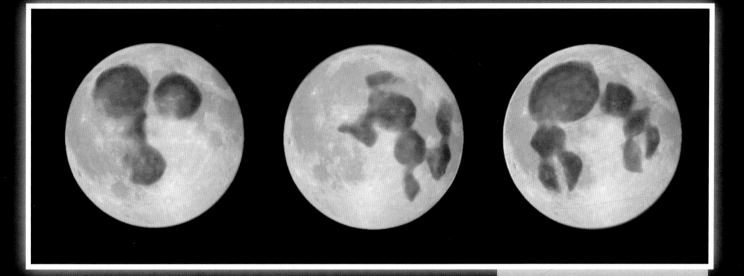

The most famous of these images is the Man-in-the-Moon, a face familiar to humans since ancient times. There is also a leaping rabbit, a crab, a poodle and a kissing couple!

Bright ideas

Some people thought that the Moon was lit from inside by a mysterious fire, but others had the right idea – that it was a solid globe, like the Earth, illuminated by sunlight. About 500 years ago, an Italian man called Leonardo da Vinci said that the Moon's dark areas were lands and the brighter areas were seas. People would have to wait for the invention of the **telescope**, 100 years later, to find out if he was right.

⇧ People sometimes see strange shapes and faces on the Moon. Above from left to right: the Man-in-the-Moon; a leaping rabbit; a crab.

An ancient Indian ⇨ myth said that the **mineral** selenite was made by moonbeams playing upon clear water. It was supposed to have remarkable healing powers.

How was the Moon made?

It was the invention of the **telescope**, around 400 years ago, that allowed **astronomers** to finally see what the Moon's surface was like. The telescope revealed that the Moon is a solid, rocky world, with large, grey plains surrounded by **highlands** packed with **craters** and mountains. Astronomers then began to wonder how the Moon was originally formed.

The Galileo **probe** took this photograph from above the Moon's north pole.

A sister planet?

Some astronomers thought that the Moon might be the Earth's sister planet, formed from the same cloud of dust and gas which created the Earth. Others thought it was so unlike the Earth that it must have formed somewhere else in the **Solar System**, and been pulled in towards the Earth by Earth's **gravity** when it passed close by.

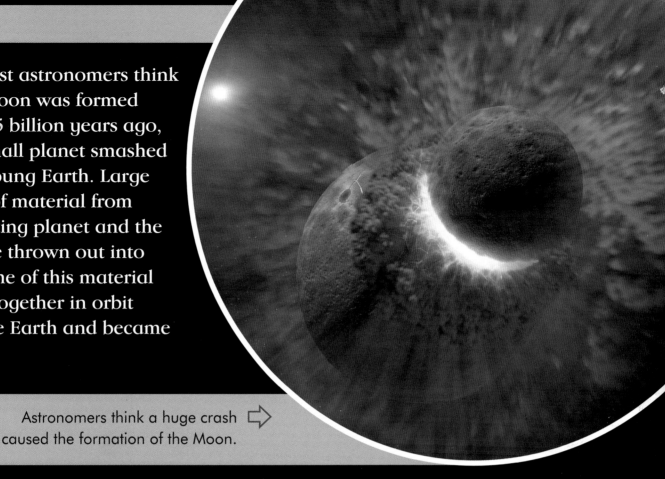

Today, most astronomers think that the Moon was formed around 4.5 billion years ago, when a small planet smashed into the young Earth. Large amounts of material from the impacting planet and the Earth were thrown out into space. Some of this material gathered together in orbit around the Earth and became the Moon.

Astronomers think a huge crash ⇨ caused the formation of the Moon.

Key Concept

Patchwork Moon

The dark patches on the Moon are made up of a dark volcanic rock similar to basalt. **On the Earth, basalt is spewed out as runny** lava **by erupting volcanoes. The Moon's brighter areas are made of rocks containing lighter minerals.**

Moonstuff

The Moon is lighter than the Earth because the Earth has a heavy iron core at its centre, and the Moon does not. At the beginning of its life, the Moon was very hot. Its has since cooled down and, unlike the Earth, is now completely solid.

This looks like ⇨ a photograph of another planet, but it is actually an area of basaltic rock on Earth's surface.

The seas of the Moon

Soon after the Moon was formed, when it was still hot inside, its surface was battered by thousands of large lumps of rock called **asteroids**. The biggest of these impacts smashed through the Moon's solid but thin rocky crust, carving out gigantic craters and allowing hot, molten rock to pour out onto the surface. These lava flows created the large, dark areas we can see, known as the Moon's **seas**. They cover a large part of the Moon's near side, but only a small part of the far side. This is because the crust on the near side was thinner than the crust on the far side, so it was easier for lava to erupt onto the surface. Eruptions of lava on the Moon stopped happening about three billion years ago, and the lava became solid.

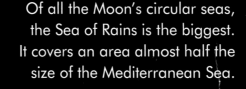

Sea of Rains

Sea of Serenity

Sea of Tranquillity

Of all the Moon's circular seas, the Sea of Rains is the biggest. It covers an area almost half the size of the Mediterranean Sea.

Sea names

Romantic names have been given to the Moon's seas, such as the Sea of Rains, the Sea of Nectar and the Ocean of Storms. At the edges of some seas are bays, such as the Bay of Rainbows and the Bay of Dew. Some dark, inland plains have also been given names, such as the Lake of Autumn and the Marsh of Sleep.

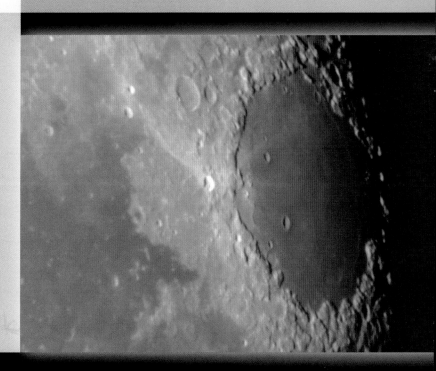

⇧ The Sea of Crises is surrounded by high mountains, seen here from above.

Lunar mountains

Huge mountain ranges surround many of the Moon's seas. Some of the mountains in ranges such as the lunar Apennines and the lunar Alps are much bigger than the Earth's own Apennines and Alps. Single mountains, or small clusters of mountains, are also found jutting up out of the seas, like islands.

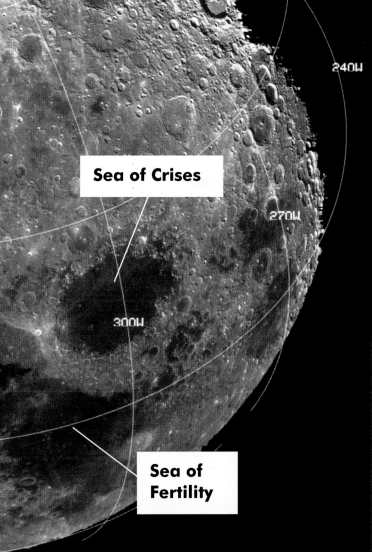

Sea of Crises

Sea of Fertility

The near side of the Moon looks different ⇨ from the far side. The near side has more dark 'seas', while the far side (shown here) is covered with craters.

Lunar craters

The Moon's highlands are covered in thousands of craters, or hollows, in the ground. Most of these were formed when asteroids hit the surface of the Moon, blasting out masses of solid rock. This happened a very long time ago. Nothing really big has hit the Moon for many millions of years.

Endless craters

While most craters look very deep, this is actually an illusion caused by the shadows cast when the Sun lights them from a low angle. Most craters are quite shallow compared to their width. If you stood inside a very large lunar crater, you might not even be able to see the walls surrounding you, because they would be beyond your horizon.

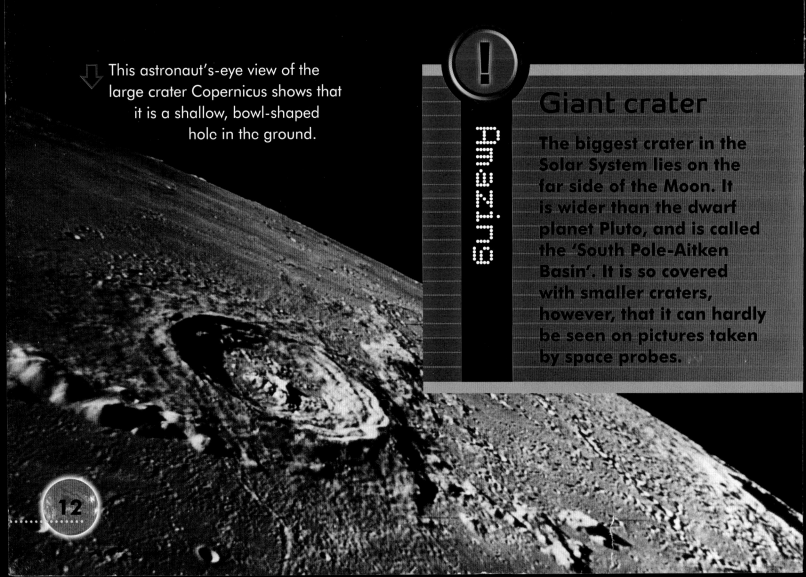

This astronaut's-eye view of the large crater Copernicus shows that it is a shallow, bowl-shaped hole in the ground.

Amazing

Giant crater

The biggest crater in the Solar System lies on the far side of the Moon. It is wider than the dwarf planet Pluto, and is called the 'South Pole-Aitken Basin'. It is so covered with smaller craters, however, that it can hardly be seen on pictures taken by space probes.

Bright ray craters

Some large craters have steep, stepped walls like the seating of a theatre. Many have large mountain peaks at their centre. Bright streaks of material, called **rays**, spread out from many of the Moon's largest young craters. This is material that was blasted out to enormous distances by the impact that formed the crater.

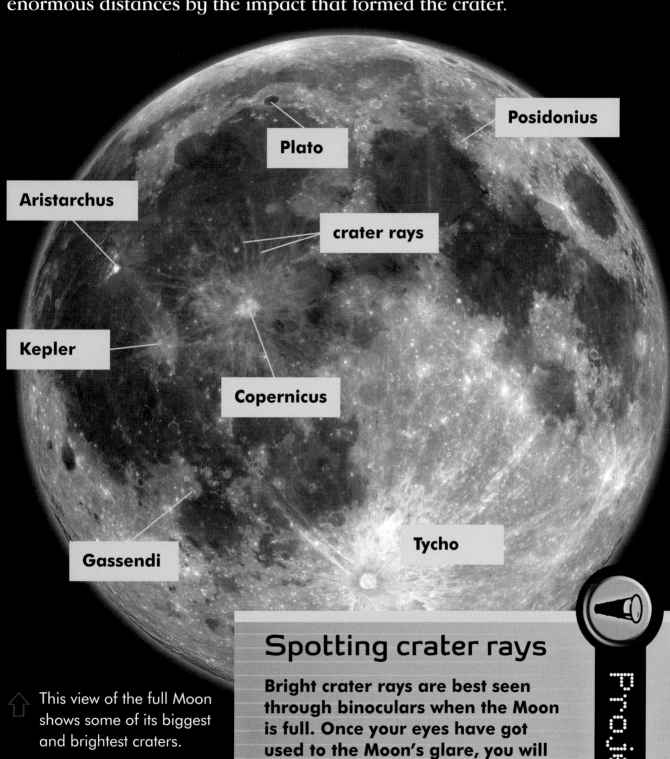

Posidonius

Plato

Aristarchus

crater rays

Kepler

Copernicus

Gassendi

Tycho

⬆ This view of the full Moon shows some of its biggest and brightest craters.

Spotting crater rays

Bright crater rays are best seen through binoculars when the Moon is full. Once your eyes have got used to the Moon's glare, you will see many bright rays. The brightest are around the crater Copernicus and around Tycho, in the Moon's southern highlands.

Project

Cracks in the crust

The Moon's crust is made of solid rock. It can stretch a little if it is pulled apart slightly by movement beneath its surface, but after a certain amount of pulling it cracks. These cracks in its crust are called **faults**, and astronomers have discovered hundreds of them on the Moon.

Lunar faults

Most lunar faults are thought to have been made in the Moon's early history, after the crust had cooled and the big asteroid impacts and volcanoes had finished sculpting its surface.

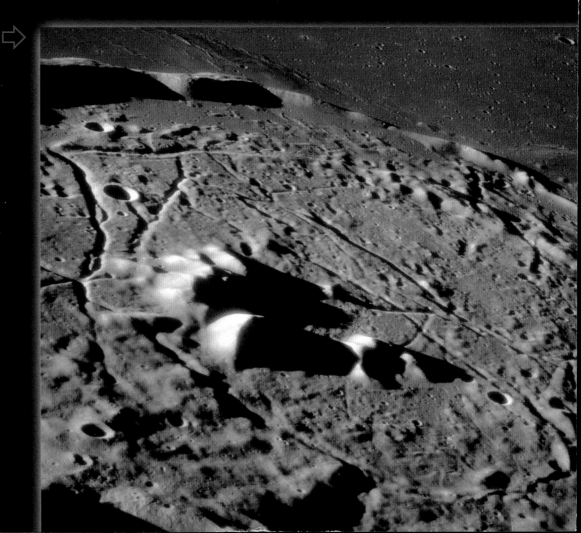

The floor of the crater Gassendi is full of **rilles** (see Fault valleys opposite). Rilles can be found inside large craters, where they form web-like networks.

Impressive faults

The neatest looking large fault visible on the Moon is called the Straight Scarp. Here, part of the Sea of Clouds has been pulled apart to create a cliff that is about 250m high and 110km long. Other lunar cliffs caused by faulting are less neat. Some distance from the edge of the Sea of Nectar stands the Altai Scarp – a giant, winding cliff around 500km long and in places up to 1000m high.

Fault valleys

In some places, the Moon's crust has been pulled apart to produce two faults lying next to each other. Where the crust between the faults has sunk down, a fault valley has been produced. Small fault valleys are known as rilles. Rilles occur around the edges of lunar seas, and also cut across highlands and craters.

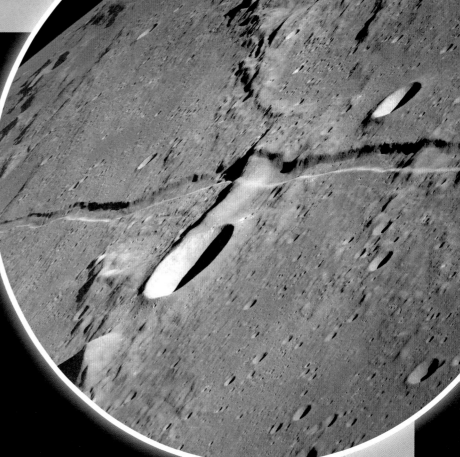

⇧ The Aridaeus Rille on the Moon cuts cleanly through all obstacles. It was formed when the Moon's crust was pulled slightly apart.

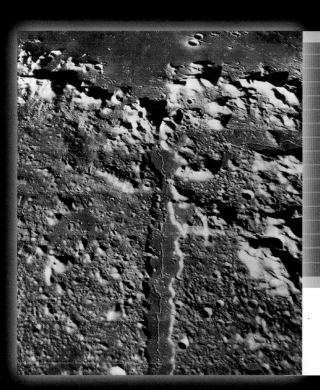

The Moon's Grand Canyon

The biggest fault valley on the Moon cuts cleanly through the Moon's Alps, as if it had been sliced out of the mountains with a gigantic chisel. The Alpine Valley is 130km long, up to 20km wide and 2000m deep.

Amazing

⇦ The Lunar Orbiter (one of **NASA**'s space probes) photographed the Alpine Valley.

The top ten lunar features

With binoculars or a small telescope, there is plenty you can see on the Moon. These are some of the most impressive features to look at:

1 COPERNICUS Often called the Monarch of the Moon, the crater Copernicus is 93km across. Its floor is more than 3500m below its rim. It has a group of central mountains, and is surrounded by bright rays.

2 ARISTARCHUS This is the Moon's brightest crater. It is 40km across, 3000m deep, and its stepped inner walls look like an Ancient Greek **amphitheatre**. A mountain peak rises from the centre.

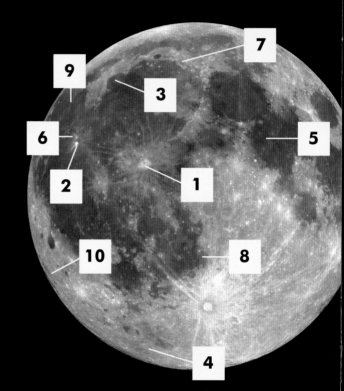

⇧ This map shows the locations of the Moon's top ten features.

⇦ Aristarchus and the nearby Schroeter's Valley.

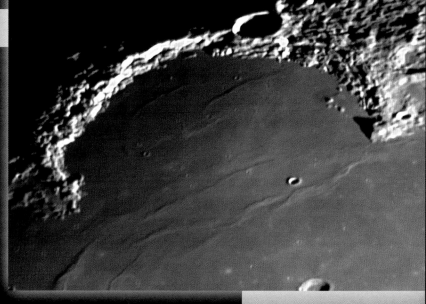

3 THE BAY OF RAINBOWS This beautiful semicircular bay on the edge of the Sea of Rains is bordered by the magnificent Jura Mountains. One headland of the bay, called Cape Heraclides, looks like the side of a woman's face, with long flowing hair behind, so it has been called the Moon Maiden.

⇧ The Bay of Rainbows.

4 WARGENTIN This unusual crater was flooded nearly to the brim with lava flows, so it looks more like a circular plateau.

5 LAMONT This is a large collection of wrinkles in the Sea of Tranquillity, surrounding a circular formation. When it is lit by a low Sun, it looks like a bullet hole in a sheet of glass.

6 SCHROETER'S VALLEY The largest lava valley on the Moon is a winding rille, 160km long and in places 1000m deep.

7 THE ALPINE VALLEY This fault valley, 180km long, cuts cleanly through the Moon's Alps.

8 THE STRAIGHT SCARP This is the so-called Straight Wall in the Sea of Clouds. It is a cliff caused by faulting in the Moon's crust.

9 MOUNT RÜMKER The biggest **dome** on the Moon is a long-extinct volcano. It is about 70km wide.

10 THE EASTERN SEA One of the Moon's most spectacular asteroid **impact craters** looks like a giant bulls-eye target. It is made up of a central lava plain, surrounded by rings of mountains and several dark lakes. Overall it measures 1000km across.

⇦ Mount Rümker.

Strange world of the Moon

For billions of years, the Moon has been affecting our planet and its inhabitants. The gravity of the Moon and the Sun pulls on the oceans of the Earth, causing the tides.

Perhaps the biggest influence the Moon has on the Earth is in causing the tides.

The beginning of life

In the very distant past, the Moon was much closer to the Earth and appeared as a huge, bright globe in the sky. Tides caused by the Moon's gravity were much stronger than they are today. Their strong movements helped to create the first life forms on Earth, in the oceans and on its shores.

In mythical stories, people can be transformed into werewolves by moonlight!

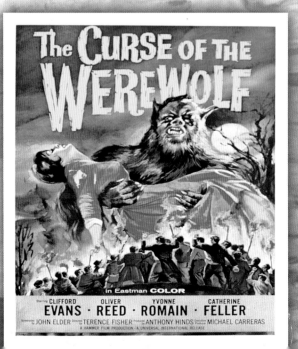

THE CURSE OF THE WEREWOLF

in Eastman COLOR

Starring CLIFFORD EVANS · OLIVER REED · YVONNE ROMAIN · CATHERINE FELLER

Screenplay by JOHN ELDER Directed by TERENCE FISHER Produced by ANTHONY HINDS Executive Producer MICHAEL CARRERAS
A HAMMER FILM PRODUCTION · A UNIVERSAL INTERNATIONAL RELEASE

A moonlit life cycle

Moonlight plays a big part in the lives of many creatures. One example is the mayfly that lives around Lake Victoria, in Africa. It hatches out during the full Moon and finds a mate. The females lay their eggs in water. Then, with their life cycle complete, the mayflies die.

Moon weather

Each year there are 12 or 13 full Moons. In ancient times, people linked these with events in their lives. For example, the full Moon in September was known as the Harvest Moon, because it was when the harvest was brought in. The ancients also thought that they could predict the weather by looking at the Moon. Sayings such as 'clear Moon, frost soon', show that people believed the weather could be predicted by looking at the Moon. In fact, the Moon has no effect at all on our weather.

⬆ The Ancient Greek astronomer Cornelius Agrippa believed that big cats are affected by the Moon. He claimed that their spots and eyes changed size with the Moon's phases. Although untrue, this idea has since become legend.

By the light of the Moon

The full Moon is bright enough to read by, but colours cannot be seen by moonlight alone. Try looking at a magazine or a colourful comic book beneath a bright Moon, away from any other lights. You may find that colours are difficult to tell apart.

19

Solar eclipses

An **eclipse** of the Sun happens when the Moon moves directly between the Earth and the Sun, blocking its light.

Partial eclipse

In a partial eclipse, only part of the Sun is covered by the Moon. At first, the curved silhouette of the Moon appears at the Sun's edge, then it gradually moves across part of its surface. Partial eclipses usually occur several times a year, but they can only be seen from certain parts of the Earth's surface. In the UK, there may be several years between them.

Total eclipse

Total eclipses are much rarer. They can only be seen from a narrow area of the Earth's surface, where the shadow of the Moon sweeps across it. The Sun is 400 times bigger than the Moon, but it is also 400 times further away, so the two objects look about the same size to us. Since the Moon can only just cover the whole Sun, a total eclipse only lasts for a few minutes.

The Sun was completely hidden by the Moon in a total solar eclipse in March 2006.

⇧ High above the
Mediterranean,
the crew of the
International Space
Station photographed
the 2006 eclipse,
looking at the Moon's
shadow on the Earth.

Pearly plumes

During a total solar eclipse, the pearly white plumes of the Sun's hot outer atmosphere **can be seen behind the Moon, along with the flame-red tongues of gas that jet off the Sun's surface.**

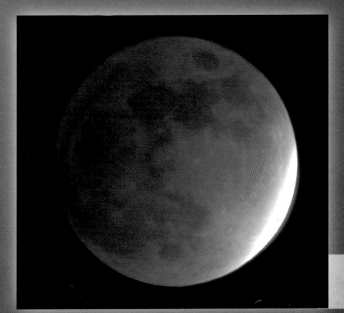

Eclipse of the Moon

Lunar eclipses are less spectacular, but wonderful in their own way. They take place when the Moon passes through the shadow cast into space by the Earth. For an hour or two, the Moon turns a lovely shade of orange or red, as sunlight is bent around the edge of the Earth onto the Moon.

 This is a total eclipse of the Moon.

Probes to the Moon

There is a limit to how much you can learn about the Moon just by observing it from the Earth. For centuries, astronomers longed to view its surface up close, to study its far side, and to know what it was made of.

Reaching the Moon

In September 1959, the Soviet Union launched the first probe to reach the Moon. A month later, another Soviet probe took the first photographs of the Moon's far side. The photos surprised astronomers – the far side had very few dark areas and many craters.

NASA's soft-landing probe ⇧
Surveyor 3 touched down on Moon's
Sea of Islands in April 1967.

Ranger, Surveyor and Lunar Orbiter

The United States responded with three very different kinds of robotic Moon mission – Ranger, Surveyor and Lunar Orbiter. The Ranger probes were one-way missions: they headed for the lunar surface, taking hundreds of pictures as they closed in, then smashed into the Moon at high speed.

The Surveyor probes touched down on the Moon, and studied the surface in detail. They found that the soil was a few centimetres thick, and able to take the weight of a manned lander. Five Lunar Orbiter probes circled the Moon, and returned thousands of pictures of its surface.

⬇ Ranger 8 took this picture of the southern part of the Sea of Tranquillity, before it crashed into the Moon in February 1965.

⬆ This picture of the Moon's mysterious far side was taken by Lunar Orbiter 3.

Russian probes

Later Russian probes brought some of the Moon's soil back to Earth. In the 1970s, two strange-looking, eight-wheeled robots called Lunokhod crawled around the Moon and took hundreds of pictures.

Space probes

Space probes allow scientists to study the Moon from up close. Orbiting probes have mapped the entire Moon, while softlanders **have examined small parts of its surface in detail. A few probes have brought samples of the Moon's soil and rock back to the Earth.**

Key Concept

Walking on the Moon

During the years 1969–1972, the USA's impressive **Apollo** programme successfully landed 12 brave astronauts on the surface of the Moon.

Saturn V

Saturn V was the vast rocket which carried the Apollo spacecraft to the Moon. It had three stages: the first stage heaved the spacecraft high above the atmosphere, the second stage pushed it into orbit and the third stage launched it towards the Moon.

Apollo 11

In July 1969, Apollo 11 carried astronauts Neil Armstrong, Buzz Aldrin and Michael Collins to the Moon. On 20 July, the landing craft Eagle touched down on the Sea of Tranquillity. Armstrong opened the hatch, climbed down the ladder and placed the first human foot on lunar soil. Shortly afterwards, Aldrin followed him. This first lunar walk lasted two and a half hours.

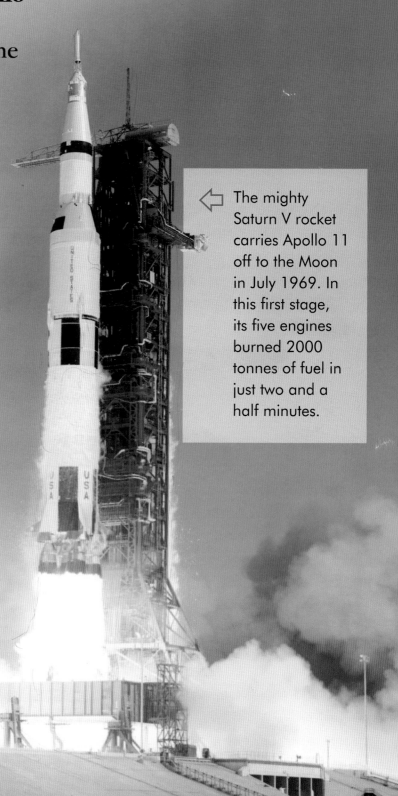

The mighty Saturn V rocket carries Apollo 11 off to the Moon in July 1969. In this first stage, its five engines burned 2000 tonnes of fuel in just two and a half minutes.

First on the Moon – Neil Armstrong (born 1930)

Neil Armstrong learned to fly before he was old enough to drive. He trained as a pilot, and became an astronaut with the Gemini Space Program. **He joined the Apollo Program and commanded the famous Apollo 11 mission. His words on being the first person to step onto the Moon were, "That's one small step for a man – one giant leap for mankind."**

⇧ Armstrong, Collins and Aldrin were the crew of Apollo 11. Collins never walked on the Moon. He orbited the Moon alone in the command module, Columbia.

⇩ During the Apollo 11 Moon walk some experiments were carried out, such as this Moonquake detector which Aldrin set up on the Sea of Tranquillity. A few boxes of soil and rock were also collected.

Apollo 12, Apollo 13 and Apollo 14

Four months later, Apollo 12's landing craft, Intrepid, landed on the Ocean of Storms, near the Surveyor space probe. The astronauts visited Surveyor, which was coated with dust. In 1970, Apollo 13 was cancelled after an explosion, but later that year Apollo 14's landing craft, Antares, touched down near the crater Fra Mauro. To the amusement of viewers on Earth, astronaut Al Shepard whacked a golf ball with a soil sampling stick.

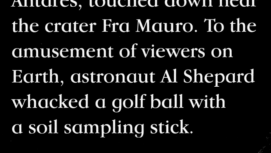

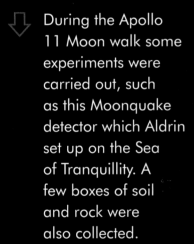

More Moon exploration

An amazing electric Moon buggy was used on the final three Apollo missions. It allowed the astronauts to explore areas several kilometres away from their landing vehicle.

Amazingly, the Moon buggy's tyres were made out of piano wire! ⇨

Apollo 15

In July 1971, the astronauts of Apollo 15 explored the shoreline of the Sea of Rains. They visited a winding valley called the Hadley Rille, and found a patch of green soil that had been sprayed out of a lava fountain. They showed that objects of different weights fall at the same speed on the airless Moon. They dropped a hammer and a falcon's feather from the same height, and both hit the Moon's surface at the same time.

Apollo 16

A hilly region called the Descartes Highlands was visited by Apollo 16 in April 1972. Using the Moon buggy, astronauts explored Stone Mountain and North Ray crater. The crater's inside walls were found to be layered, where ancient lava flows had built up over time.

Apollo 17

Apollo 17, the last mission to the Moon, visited an area near the crater Littrow in December 1972. It discovered some bright orange soil, which was later found to be made of tiny coloured beads formed in the intense heat of a meteorite impact.

Soviet space projects

The Soviet Union also had a lunar orbiting and landing programme. They developed techniques useful for a Moon landing and they sent unmanned probes to orbit and land on the Moon. Their mighty rocket, the N-1, was just as powerful as Saturn V, but it failed disastrously. In the end, their expensive project was abandoned when it became clear that the United States was winning the race to put people on the Moon.

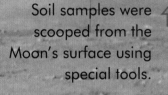

Soil samples were scooped from the Moon's surface using special tools.

Apollo 16 astronaut ⇨ John Young leaps above the lunar surface as he salutes the US flag.

The future Moon

In recent years, several unmanned lunar probes have been sent into space, including the United States' Clementine and Lunar Prospector missions, and the European SMART-1 mission. They have all helped to map the Moon in great detail, and have discovered more about the materials that make up the Moon's surface.

The Crew Exploration Vehicle will orbit the Moon while astronauts explore its surface.

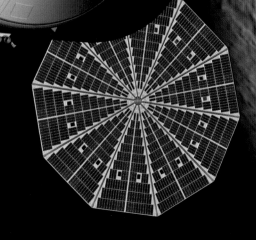

People return to the Moon

The USA's space agency, NASA, said that it plans to return astronauts to the Moon by 2020. Four astronauts will ride into orbit in a Crew Exploration Vehicle, twice the size of the old Apollo command module. In orbit it will **dock** with an unmanned lunar lander. Both vehicles will dock with a large booster, to make the two-day outbound voyage to the Moon.

Once in orbit around the Moon, the astronauts will put the Crew Exploration Vehicle on to autopilot, and make their way down to the Moon's surface in the lander. They will spend seven days on the Moon, carrying out experiments, making observations and exploring their surroundings.

All astronauts visiting the Moon will leave behind equipment and supplies that can be used by later missions, as well as material that could be used to make a permanent Moon base.

⇧ This shows how, in orbit around the Moon, the Crew Exploration Vehicle will dock with the lunar lander.

⇦ NASA is planning to set up a permanent Moon base by 2024.

Where next?

Our return to the Moon will be a stepping stone for our further exploration of the Solar System. After the Moon will come the planet Mars. The first human footprint on that distant red planet is likely to be made in your lifetime – perhaps you will be the wearer of that first boot to tread on Mars!

Glossary

amphitheatre an oval, circular or semicircular-shaped stadium

Apollo the Moon landing missions of the United States

asteroid a big lump of rock in space. When an asteroid hits the Moon it makes a crater

astronomer someone who studies astronomy – the scientific study of objects in space

atmosphere the mixture of gases surrounding a star or a planet

basalt a dark volcanic rock

crater a circular, bowl-shaped pit blasted out of a solid surface when it is hit by an asteroid. Volcanoes also have small craters at their tops

crescent the shape of the Moon at the start and end of a lunar month

dock to join up with another spacecraft in space

dome a low, rounded lunar volcano

earthshine the faint blue-tinted glow of the Moon's unlit area, visible with the naked eye when the Moon is a narrow crescent. It is caused by sunlight reflected onto the Moon by the Earth

eclipse when the Moon moves directly between the Earth and the Sun it blocks out the Sun's light and causes a solar eclipse. When the Moon moves into the shadow of the Earth, it produces a lunar eclipse

fault a crack in the Moon's crust

full Moon when the Moon appears fully lit up by the Sun

Gemini Space Program NASA's 10 manned missions that launched two-person crews into orbit to try out procedures that would be useful in a Moon landing

gravity a pulling force that acts throughout the Universe. The bigger and more dense an object, the more gravity it has

highlands heavily cratered or mountainous areas on the Moon. They appear much brighter than the seas

impact crater a pit in the Moon's surface formed by an object hitting the Moon at high speed

lava hot, melted rock which has bubbled up from below a planet's surface (usually through a volcano)

lunar an adjective that describes anything to do with the Moon

Moon the Moon is the Earth's only natural satellite. The satellites of other planets are also known as moons (with a small 'm')

mineral a solid material that is found naturally in the ground. Rocks are made out of minerals

NASA the United States' national space agency. The letters stand for 'National Aeronautics and Space Administration'

orbit the curved path of a planet or other object round a star, or a moon round a planet

phase the amount by which the Moon appears lit up by the Sun

planet a large, rounded object orbiting a star. The Sun has eight main planets – Mercury, Venus, Earth, Mars, Jupiter, Saturn, Uranus and Neptune. Pluto is a dwarf planet

probe an automatically controlled spacecraft which gathers scientific information

ray a bright line streaking away from an impact crater across the Moon's surface. Rays are made up of material thrown out by the impact that made the crater

rille a narrow lunar valley caused by faulting as the Moon's crust has pulled apart. They can be straight or curved. Others are winding and snake-like, thought to have been caused by fast-moving lava flows

satellite a small object in orbit around a larger object. The Moon is a satellite of the Earth

seas large plains of basaltic lava on the Moon, that look darker than the surrounding areas

softlanders space probes that gently land upon the surface of the Moon or planet

Solar System our part of the Universe, containing the Sun, the planets and their moons, dwarf planets, asteroids and comets

space everything outside the Earth

Sun our nearest star. It is a huge, exploding ball of gas

telescope an instrument used by astronomers to study objects in space. Telescopes can show the Moon's mountains, craters and seas very well

Index